Neighborhood Safari

Bats

by Martha London

www.focusreaders.com

Copyright © 2021 by Focus Readers®, Lake Elmo, MN 55042. All rights reserved. No part of this book may be reproduced or utilized in any form or by any means without written permission from the publisher.

Focus Readers is distributed by North Star Editions:
sales@northstareditions.com | 888-417-0195

Produced for Focus Readers by Red Line Editorial.

Photographs ©: Shutterstock Images, cover, 1, 4, 7, 8, 11, 12, 15, 17, 18, 21 (bats); Red Line Editorial, 21 (chart)

Library of Congress Cataloging-in-Publication Data
Names: London, Martha, author.
Title: Bats / by Martha London.
Description: Lake Elmo, MN : Focus Readers, [2021] | Series: Neighborhood safari | Includes index. | Audience: Grades 2-3
Identifiers: LCCN 2019060193 (print) | LCCN 2019060194 (ebook) | ISBN 9781644933503 (hardcover) | ISBN 9781644934265 (paperback) | ISBN 9781644935781 (pdf) | ISBN 9781644935026 (ebook)
Subjects: LCSH: Bats--Juvenile literature.
Classification: LCC QL737.C5 L66 2021 (print) | LCC QL737.C5 (ebook) | DDC 599.4--dc23
LC record available at https://lccn.loc.gov/2019060193
LC ebook record available at https://lccn.loc.gov/2019060194

Printed in the United States of America
Mankato, MN
082020

About the Author

Martha London writes books for young readers. When she's not writing, you can find her hiking in the woods.

Table of Contents

CHAPTER 1
Flying at Night 5

CHAPTER 2
Mammals with Wings 9

CHAPTER 3
Strong Senses 13

THAT'S AMAZING!
Echolocation 16

CHAPTER 4
Life in a Group 19

Focus on Bats • 22
Glossary • 23
To Learn More • 24
Index • 24

Chapter 1

Flying at Night

A bat swoops across the night sky. It is hunting. The bat flaps its wings. Then it grabs a moth in its mouth.

Bats live in many countries around the world. Bats often live in caves or trees. Sometimes, bats **roost** in the attics of houses. These dark places protect bats from **predators**.

Fun Fact There are hundreds of different **species** of bats.

Chapter 2

Mammals with Wings

A bat's body is covered in fur. The fur is often gray, red, or brown. Bats have sharp teeth. Many bats have large ears.

Bats are the only **mammals** that fly. Some bats' wings spread 6 feet (1.8 m) across. Other bats are only a few inches wide.

A bat's wings have thin skin. The skin goes over long bones.

Fun Fact

The world's smallest bat is the size of a thumbnail.

Chapter 3

Strong Senses

Bats use their senses to stay safe. Bats have good hearing. They listen for predators such as hawks and snakes.

Bats also use their senses to find food. Some bats eat fruit. Most bats eat insects. Bats listen for the sounds the insects make. Many bats have good eyesight, too. They can see in the dark.

Fun Fact: Bats can eat more than 1,000 mosquitoes every hour.

That's Amazing!

Echolocation

Many bats use **echolocation** to find food. Bats make high-pitched sounds. The sound waves bounce off nearby objects. They bounce back to the bats. Bats sense these echoes. The echoes create a kind of picture for bats. Bats use it to know where objects are.

Chapter 4

Life in a Group

Most bats live in large groups. These groups are called colonies. The bats in a colony work together to hunt and to raise babies.

Some colonies have just a few bats. Others have millions. The bats rest during the day. They hang upside down. Then they fly out at night. Bats help keep insect numbers under control. They also help **pollinate** plants.

Life Cycle

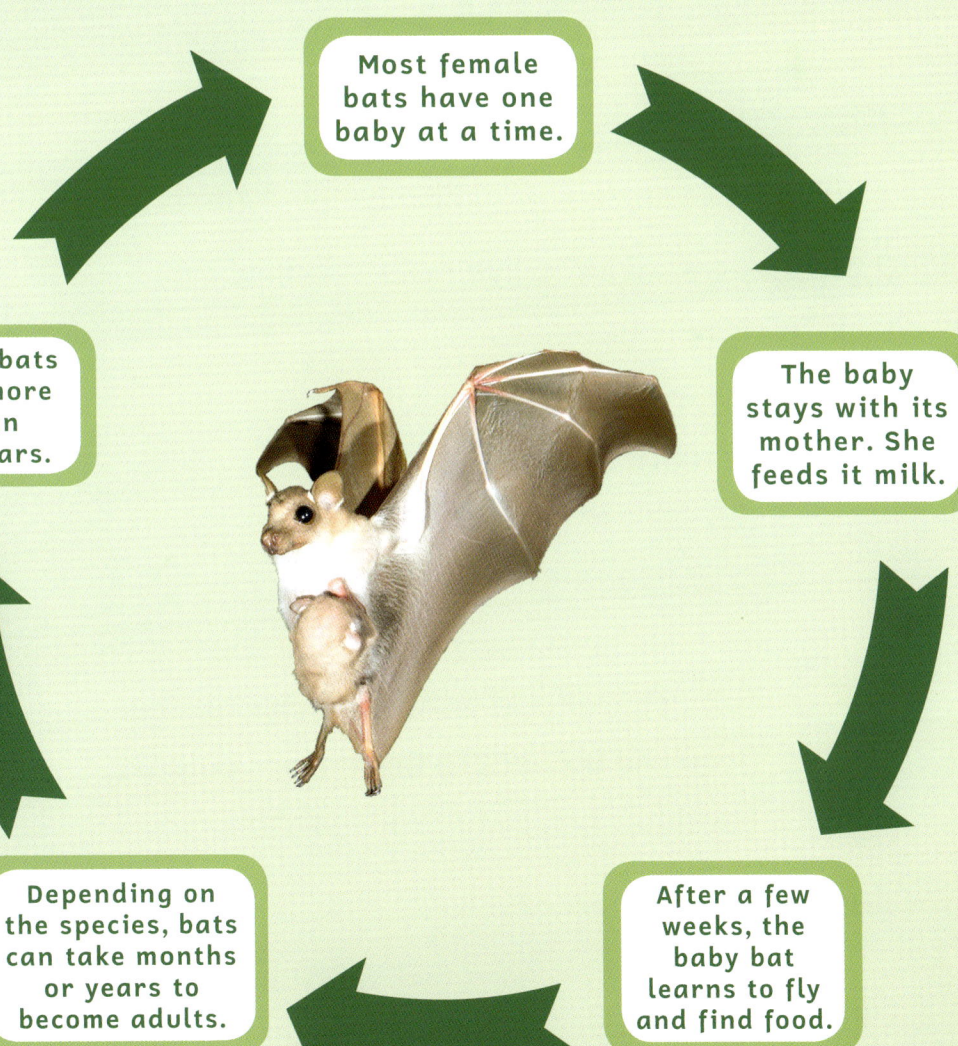

Most female bats have one baby at a time.

The baby stays with its mother. She feeds it milk.

After a few weeks, the baby bat learns to fly and find food.

Depending on the species, bats can take months or years to become adults.

Some bats live more than 30 years.

21

FOCUS ON
Bats

Write your answers on a separate piece of paper.

1. Write a sentence describing how bats hunt.
2. Would you want to live near a bat colony? Why or why not?
3. When do bats come out to hunt?
   - A. at night
   - B. during the day
   - C. only in winter
4. How do bats help keep insect numbers low?
   - A. Bats share food with the insects.
   - B. Bats use insects to build their colonies.
   - C. Each bat can eat many insects.

Answer key on page 24.

Glossary

echolocation
A method of using sound waves to sense where objects are located.

mammals
Animals that have hair and feed their babies milk.

pollinate
To spread pollen from plant to plant. Spreading pollen helps plants grow seeds.

predators
Animals that hunt other animals for food.

roost
To settle somewhere to rest or sleep.

species
A group of animals or plants of the same kind.

To Learn More

BOOKS

Bodden, Valerie. *Bats*. Mankato, MN: Creative Paperbacks, 2020.

Samuelson, Benjamin O. *Journey of the Bats*. New York: Gareth Stevens Publishing, 2018.

NOTE TO EDUCATORS

Visit **www.focusreaders.com** to find lesson plans, activities, links, and other resources related to this title.

Index

C
colonies, 19, 20

E
echolocation, 16

F
food, 14, 16, 21

P
predators, 6, 13

S
senses, 13, 14, 16

W
wings, 5, 10

Answer Key: 1. Answers will vary; 2. Answers will vary; 3. A; 4. C